Community Places

Fire Station

by Amy McDonald

Blastoff! Beginners

BELLWETHER MEDIA
MINNEAPOLIS, MN

Blastoff! Beginners are developed by literacy experts and educators to meet the needs of early readers. These engaging informational texts support young children as they begin reading about their world. Through simple language and high frequency words paired with crisp, colorful photos, Blastoff! Beginners launch young readers into the universe of independent reading.

Sight Words in This Book

a	eat	the
and	here	they
are	is	this
at	look	to
big	out	we

This edition first published in 2022 by Bellwether Media, Inc.

Library of Congress Cataloging-in-Publication Data

Names: McDonald, Amy, 1985- author.
Title: Fire station / by Amy McDonald.
Description: Minneapolis, MN : Bellwether Media, Inc., 2022. | Series: Blastoff! beginners : Community places | Includes bibliographical references and index. | Audience: Ages PreK-2 | Audience: Grades K-1
Identifiers: LCCN 2021044266 (print) | LCCN 2021044267 (ebook) | ISBN 9781644875650 (library binding) | ISBN 9781648346569 (paperback) | ISBN 9781648345760 (ebook)
Subjects: LCSH: Fire stations--Juvenile literature.
Classification: LCC TH9148 .M3295 2022 (print) | LCC TH9148 (ebook) | DDC 628.9/25--dc23
LC record available at https://lccn.loc.gov/2021044266
LC ebook record available at https://lccn.loc.gov/2021044267

Editor: Christina Leaf Designer: Andrea Schneider

Printed in the United States of America, North Mankato, MN.

Table of Contents

At the Fire Station! 4
What Are Fire Stations? 6
Life at the Fire Station 8
Fire Station Facts 22
Glossary 23
To Learn More 24
Index 24

At the Fire Station!

Look at the truck!
We are at
the fire station.

What Are Fire Stations?

Fire stations are big places. They hold fire trucks.

fire truck

17
PORTLAND
21

Life at the Fire Station

Firefighters work
at the station.
They wait
to fight fires.

firefighters

This is the kitchen. Firefighters cook and eat here.

ENGINEER

This is the gym. They work out to stay strong.

RAID AVENTURE

This is the **dorm**. They sleep at the fire station.

dorm

A bell rings!
Firefighters hurry
to the **garage**.

OK & LADDER CO. 5
ENGINE CO
REMEMBER
OUR FALLEN BROTHERS
FDNY
E-24
L-5
F.D.
N.Y.
5
F.D.N.Y.
GMC

garage

They need **gear**. **Lockers** hold helmets and jackets.

helmets
lockers
jackets

Open the doors!
We are off
to a fire!

NEVER MISSED A PERFORMANCE
54
F.D.
N.Y.
54

Fire Station Facts

Inside a Fire Station

What Happens in a Fire Station?

firefighters eat

firefighters work out

firefighters get gear

Glossary

dorm

a shared bedroom in a fire station

garage

a place to park cars and trucks

gear

things firefighters use to fight fires

lockers

places that hold gear for firefighters

To Learn More

ON THE WEB

FACTSURFER

Factsurfer.com gives you a safe, fun way to find more information.

1. Go to www.factsurfer.com.
2. Enter "fire station" into the search box and click 🔍.
3. Select your book cover to see a list of related content.

Index

bell, 16
cook, 10
dorm, 14
eat, 10
fire truck, 4, 6
firefighters, 8, 9, 10
fires, 8, 20
garage, 16, 17
gear, 18
gym, 12
helmets, 18, 19
jackets, 18, 19
kitchen, 10
lockers, 18, 19
sleep, 14
work, 8
work out, 12

The images in this book are reproduced through the courtesy of: Peter Fleming, front cover (station); ryasick, front cover (truck); Flashon Studio, pp. 3, 18; FrankvandenBergh, pp. 4-5; Le Do, p. 6 (truck); zhudifeng, pp. 6-7; PrasongTakham, pp. 8-9 (fire); Lpettet, pp. 8-9 (firefighters); stockcreations, p. 10 (chili); dlewis33; pp. 10-11; Happy Stock Photo, p. 12; Forget Patrick/ Alamy, pp. 12-13, 22 (work out); The Color Archives/ Alamy, p. 14; FirePhoto/ Alamy, pp. 14-15; Michael Gordon, pp. 16-17; Christa Eder, pp. 18-19; Manuel Esteban, pp. 20-21; Pabkov, p. 22 (inside a fire station); Monkey Business Images, p. 22 (eat); Tyler Olson, pp. 22 (get gear), 23 (gear); Mayuree Moonhirun, p. 23 (dorm); Colin Temple, p. 23 (garage); Rasto Blasko, p. 23 (lockers).